A TRUE BOOK™

UNDERSTANDING CLIMATE CHANGE

Animals in Danger

Katie Free

Children's Press®
An Imprint of Scholastic Inc.

Content Consultant
Heidi A. Roop, PhD
Research Scientist
Climate Impacts Group
University of Washington, Seattle
Seattle, Washington

Library of Congress Cataloging-in-Publication Data
Names: Free, Katie, author.
Title: Animals in danger: understanding climate change/Katie Free.
Other titles: True book.
Description: New York: Children's Press, an imprint of Scholastic Inc., 2020 | Series: A true book | Includes
index. | Audience: Grades 4–6. | Summary: "Book describes climate change and how it affects the
animals"—Provided by publisher.
Identifiers: LCCN 2019031392 | ISBN 9780531130759 (library binding) | ISBN 9780531133750 (paperback)
Subjects: LCSH: Endangered species—Juvenile literature. | Climatic changes—Juvenile literature. | Nature—
Effect of human beings on—Juvenile literature.
Classification: LCC QL83 .F74 2020 | DDC 591.68—dc23
LC record available at https://lccn.loc.gov/2019031392

Design by THREE DOGS DESIGN LLC
Produced by Spooky Cheetah Press
Editorial development by Mara Grunbaum

Scholastic Inc., 557 Broadway, New York, NY 10012

1 2 3 4 5 6 7 8 9 10 R 29 28 27 26 25 24 23 22 21 20

Front cover: A polar bear navigates melting sea ice.

Back cover: Coral reefs are at risk because of climate change.

Find the Truth!

Everything you are about to read is true *except* for one of the sentences on this page.

Which one is **TRUE**?

T or F Climate change will significantly decrease mosquito populations.

T or F Climate change is making it hard for ocean life to grow shells.

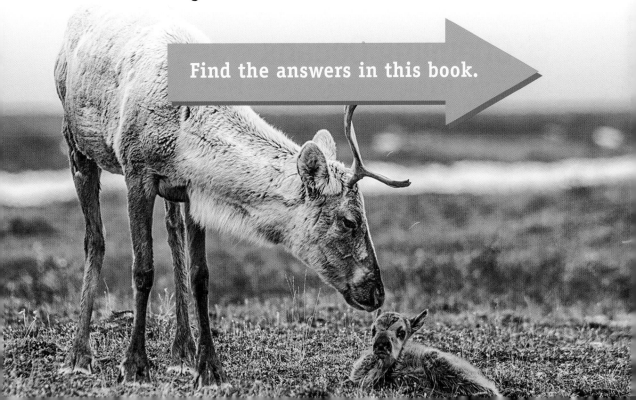

Find the answers in this book.

Contents

Bees are
sensitive to
climate change.

Young people
can play a role in
protecting local
wildlife populations.

A Critical Moment

In recent centuries, humans have released increasing amounts of **greenhouse gases** into Earth's **atmosphere**. These gases have trapped heat in the atmosphere, causing the average temperature on the planet's surface to rise and contributing to **global warming**.

Because of global warming, oceans are heating up, sea levels are rising, and weather is becoming more extreme. These changes in Earth's climate are known as global **climate change**. They threaten people and other plant and animal species around the world. If we don't make changes to reduce greenhouse gas **emissions** now, these problems will worsen.

There is good news, though!

Thousands of scientists around the world are studying global climate change. Politicians, public figures, and citizens of all ages are trying to figure out what to do. Humanity now knows more than ever about the causes and effects of climate change, as well as how we might reduce its impact. That means **people today can make decisions** that will affect the planet for centuries to come.

Turn the page to learn how climate change is affecting the wild animals that share our Earth.

Experts say about 8 million plant and animal species live on Earth.

Diversity at Risk

Gentoo penguins waddle along in Antarctica, the coldest and driest continent on the planet. Flamingos, wildebeests, and gazelles roam the warm African savanna. Colorful fish dart in and out of warm-water coral reefs. Monarch butterflies migrate thousands of miles every fall to escape the cold weather. Animals thrive in different environments around the world—and each creature plays a role in its **ecosystem**.

A Fine Balance

The animals that live together keep their ecosystem in balance. If a key animal is removed, other species can suffer. For example, by the mid-1920s there were no wolves left in Yellowstone National Park. They had been wiped out by hunters. But wolves were top predators in the park's ecosystem. As a consequence, more elk survived and ate young trees. That caused forests to shrink, along with bird populations.

Before

After

This image shows Yellowstone before (left) and after (right) the reintroduction of wolves in 1995.

Red-and-green macaws are just one of 800 bird species in Manu National Park.

A Healthy Mix

When an ecosystem has more plant and animal species, it has greater **biodiversity**. Habitats with higher biodiversity tend to be more resilient to change.

Manu National Park in Peru, South America, is a biodiversity hot spot. More than 1,000 animal species are found there.

They are more able to weather natural disasters such as storms and wildfires. They are also better equipped to survive disease outbreaks.

Out of Whack

Unfortunately, human activity can change ecosystems at a very fast rate. We cut down forests to build homes, harvest logs, and plant crops. We pollute landscapes with litter, and we release chemicals into waterways. These changes harm wildlife and damage habitats. Species are unable to adapt or find suitable habitats quickly enough, which reduces biodiversity.

Almost 300,000 square miles (777,000 square kilometers) of forest have been cut from the Amazon rain forest since 1970.

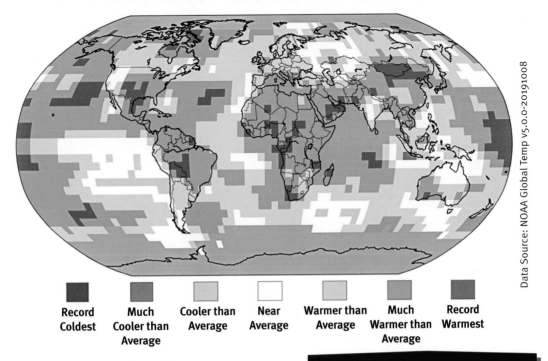

Record Coldest	**Much Cooler than Average**	**Cooler than Average**	**Near Average**	**Warmer than Average**	**Much Warmer than Average**	**Record Warmest**

Data Source: NOAA Global Temp v5.0.0-20191008

An Added Stress

This map shows the changes in average global temperatures for September 2019.

Humans are also causing ocean and air temperatures to rise. This is happening in large part because we're burning **fossil fuels** that release an excess of heat-trapping greenhouse gases into the air. One of these gases is carbon dioxide (CO_2), which is produced by cars and power plants. The more we warm the atmosphere, the more we harm the plants and wild animals that share our planet.

Coal-fired factories are the largest contributors to atmospheric CO_2.

Predators would have a hard time spotting the moth on this tree.

A Shifting Balance

In the 1800s, people in the United Kingdom burned so much coal that trees turned black with soot. Birds could easily spot and eat light-colored peppered moths on the dark trees. But dark peppered moths blended in with the soot on the trees and lived to lay eggs. As a result, peppered moth populations grew darker over time, and the species survived. The peppered moths were able to adapt quickly to a human-caused change.

Adapt to Survive

Even before humans roamed Earth, many animals had adapted or made changes to fit their environments. For example, the fennec fox lives in a hot desert, and it loses heat through its ears. Over time, the fox evolved to have large ears that help it cool down quickly. No matter the cause, animals that are able to adapt to their surroundings are more likely to survive and pass their traits along to their offspring.

Timeline of Mass Extinctions:

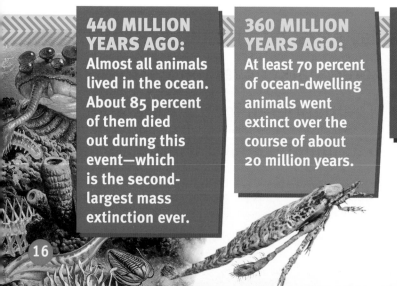

440 MILLION YEARS AGO: Almost all animals lived in the ocean. About 85 percent of them died out during this event—which is the second-largest mass extinction ever.

360 MILLION YEARS AGO: At least 70 percent of ocean-dwelling animals went extinct over the course of about 20 million years.

250 MILLION YEARS AGO: The deadliest known extinction occurred. More than 95 percent of marine species and 70 percent of land species were killed off.

Going Extinct

When a species can't adapt or find a suitable habitat, it may go extinct. This has been happening for millions of years. At times, the extinctions of large numbers of species have happened in a short period. There have been at least five of these mass extinctions. The first was when massive glaciation froze much of Earth's water. Huge lava flows contributed to the next three extinctions, and the most recent was caused by an asteroid strike.

200 MILLION YEARS AGO: More than 75 percent of sea and land animals died off. This extinction left room for dinosaurs to evolve and thrive.

65 MILLION YEARS AGO: The dinosaurs became extinct, allowing mammals to evolve.

TODAY: Many scientists think human activities—especially those that have contributed to climate change—are causing a mass extinction.

A Warming World

If Earth has gone through several mass extinctions, what is different now? This is the first time a mass extinction is being triggered by human activity. And it's happening quickly. Earth's average surface temperature has increased about 1.8 degrees Fahrenheit (1 degree Celsius) since the 1880s. Scientists estimate that the global temperature will rise between 2.5°F and 10°F (1.4°C and 5.6°C) over the next century. They estimate that 5 to 16 percent of all plant and animal species will die out if that happens.

Half the population of reindeer has been lost in the past 20 years because climate change is causing lichen, their main food source, to disappear.

Thriving with Climate Change

As the planet warms, some animal species may be able to expand into new habitats. For example, mice are very adaptable, have broad diets, and breed quickly. Jellyfish can survive in warmer water. Ticks, mosquitoes, and bowhead whales are also among the animals that might thrive in a warming world. However, this might not be good news for Earth's biodiversity. These animal species might have negative effects on the rest of the ecosystems already in place, where a balance of organisms is essential and delicate.

Because of climate change, the habitat of the American bullfrog has already expanded dramatically, harming many ecosystems.

A female sea turtle returns to the same beach every year—the one where she was born—to lay her eggs.

Hatchlings crawl from their nests to the relative safety of the sea.

Changing Oceans

As the planet warms, the ocean warms and expands and ice melts, causing sea levels to rise. This puts animals that breed on coastlines at risk. Sea turtles, for example, dig holes in beaches and lay their eggs in nests they bury in the sand. If rising waters wash away the beaches, the turtles may not be able to breed. And if more waves batter the shore, the ocean can wash away existing nests or destroy turtle eggs.

Walruses also rely on sea ice. They use it for resting and giving birth.

Melting Sea Ice

The Arctic is warming at a rate almost twice the global average, and ice covering the sea near the North Pole is shrinking. This is causing problems for Arctic animals such as ringed seals, which dig into sea ice and snow to make dens, where they raise their pups. When the ice collapses, few seal pups survive. This hurts polar bears that depend on seals for food. It also hurts people who need seals and bears for food—and who celebrate the animals as part of their culture.

Acidic Waters

Climate change also affects ocean chemistry. Right now, the ocean absorbs 22 million tons of carbon dioxide each day from the atmosphere. When this gas dissolves in seawater, the water becomes more **acidic**. As a result, animals that have shells find it harder to develop properly. This puts corals, clams, and oysters at risk. It also affects pteropods, which are sea snails and sea slugs, a major food source for whales and salmon.

Oyster farmers in Washington State are feeling the effects of climate change.

Warming Oceans

Oceans have gotten warmer by absorbing heat from the atmosphere, and the heated water is hurting many animals. Cold-water fish such as salmon and cod have a harder time breathing in warmer water. And they need to eat more food than may be available in order to survive. Heat waves that devastate ocean animals, from tiny krill to massive whales, are also becoming more common.

Overfishing by humans is also putting salmon at risk.

The Rain Forests of the Sea

People call coral reefs the rain forests of the sea because of their biodiversity. These habitats are home to a quarter of all marine life. But coral reefs are under threat from climate change. When waters get too hot for corals, the algae that live inside them leave. This can kill the corals or make them more vulnerable to disease. Researchers estimate that 2.7°F (1.5°C) of warming could cause 70 to 90 percent of corals to die. That would also put hundreds of species that depend on coral reefs, such as clown fish, sea stars, and sharks, at risk.

Corals without algae turn white—a phenomenon known as coral bleaching.

Snowshoe hares in Wisconsin are moving to snowier habitats—which is upsetting the balance in their ecosystem.

This snowshoe hare is easy to spot in a rocky landscape. The hare would be camouflaged if there were snow.

Animals on Land

Like peppered moths, snowshoe hares depend on camouflage, or the ability to blend in, to stay safe from predators. As fall turns to winter, these mammals turn from brown to white to blend in with their snowy habitat. Once spring comes, they change back to brown. But what happens when spring comes earlier because of climate change? If the hares can't adapt to change their coats sooner, predators will easily spot them, and they will be eaten. Their populations will likely diminish.

On the Tundra

The Arctic tundra is a cold and treeless area in the Far North that is warming at an alarming rate. And that spells trouble for many of its inhabitants, including arctic foxes. Arctic foxes never had to share their territory with red foxes. But as the red foxes' habitat has warmed, they've started moving north. The larger, more dominant red fox can chase arctic foxes from their territory or even kill them.

Red foxes (top) compete with arctic foxes (bottom) for limited food resources.

Out of Options

Like the red fox, some other creatures can change their range to escape the effects of climate change. But what happens

Bamboo makes up 99 percent of a giant panda's diet.

to animals that don't have that option? For example, giant pandas live in the forests of central Asia. They might be able to move to northern latitudes if their home range gets too warm, but bamboo—their main food source— might not be available in those areas.

Timing Is Everything

Shorter winters and earlier springs are causing trouble for animals whose lives follow a seasonal pattern. For example, bees drink nectar from blooming flowers. As the climate changes, the timing of flower blooms and bee life cycles may be out of sync. If that happens, bees may go hungry. Migrating birds such as the red knot may starve by showing up to breeding grounds too late, after their food source—insects—has already come and gone.

Red knots travel about 19,000 miles (30,578 kilometers) during their annual migration.

Where's the Water?

Animals may struggle as climate change disrupts rain patterns, causing **drought** or flooding in areas around the world. For example, lesser flamingos build their nests in the middle of African lakes every few years, when water levels are just right. If the water is too high, the nests will flood. If the water is too low, predators will reach the nests and eat the eggs.

If there is no rain, flamingos won't mate.

Forest Fires

As hotter and drier conditions become more common, so do wildfires. Wildfires often have natural causes, and forest animals have ways to deal with them—birds can fly away and small mammals can seek shelter, for example. But fires can be devastating to animal populations that are already small in number. After a 2017 fire in Arizona, in the area where the Mount Graham red squirrel lives, only 35 of these rare mammals were left.

Fires allow new plant life to grow on the forest floor.

Invaders

In the Florida wetlands, a Burmese python wraps its body around a raccoon, squeezes it to death, and swallows it whole. This can seem like nature in action. But Burmese pythons are an invasive species, which means they do not belong where they're living. Snakes that were brought to the United States as pets got away and began multiplying. Now they are killing native mammals such as raccoons. This is yet another example of how other human activities besides climate change disrupt natural ecosystems.

Gone Forever

If greenhouse gas emissions continue at the current rate, many species will die out. But it won't be the first time human activity has caused extinction. Hundreds of species have been wiped out by humans since the 1500s. Take a look at some of them:

NAME OF ANIMAL: Dodo
YEAR AND PLACE OF EXTINCTION:
1681; the Indian Ocean island of Mauritius
CAUSE OF EXTINCTION:
Overhunting and the introduction of predatory animals

NAME OF ANIMAL:
Caspian tiger
YEAR AND PLACE OF EXTINCTION: 1970; Turkey
CAUSE OF EXTINCTION:
Overhunting

NAME OF ANIMAL:
Bramble Cay melomys
YEAR AND PLACE OF EXTINCTION: 2016; Australia
CAUSE OF EXTINCTION: Rising sea levels

NAME OF ANIMAL:
Pinta Island tortoise
YEAR AND PLACE OF EXTINCTION: 2012; Galápagos National Park in Ecuador
CAUSE OF EXTINCTION: Overhunting and loss of habitat

NAME OF ANIMAL: **Caribbean monk seal**
YEAR AND PLACE OF EXTINCTION: 2008; the Caribbean Sea and Gulf of Mexico
CAUSE OF EXTINCTION: Overhunting

One out of every three bites of food is made possible by bees and other pollinators.

Climate change makes bees more susceptible to diseases that thrive in heat.

5

Fighting Extinction

Wild animals don't only keep delicate ecosystems in balance—they also provide benefits to people. For example, **pollinators** such as bees are important for growing fruits and vegetables. Without them, our food supply would be at risk. If we drive more animals to extinction, the consequences could be catastrophic. Scientists agree that stopping global climate change is our best hope for preserving biodiversity.

As top predators, jaguars play an important role in their ecosystem.

Protecting Ecosystems

One thing we can do is protect forests and other habitats from human activities such as logging. Trees and other plants absorb carbon dioxide out of the air. Protecting forests can improve biodiversity and help reduce climate change.

Scientists are tracking how ecosystems are adapting to climate change. This will help them identify which plants and animals are in trouble and how to work to save them.

Restoring Nature

Koalas in Australia spend most of their time in eucalyptus trees. But as weather patterns change, the trees are having a harder time surviving. Today, because of people's votes and actions, lawmakers in that nation are working to preserve koala habitats.

This is just one example of how people are taking steps to actively respect and protect the animals that share the planet with us. The more people join in, the more chances we will have to preserve our Earth's biodiversity. 🌍

Koalas can't adapt quickly enough to change.

Animals on the Edge

Scientists warn that a number of species are at risk for extinction in the next few decades due to human-made causes such as pollution, habitat loss, and climate change.

The bar graph on the next page shows the percentage of species that each group of animals may lose. Study the graph, and then answer the questions.

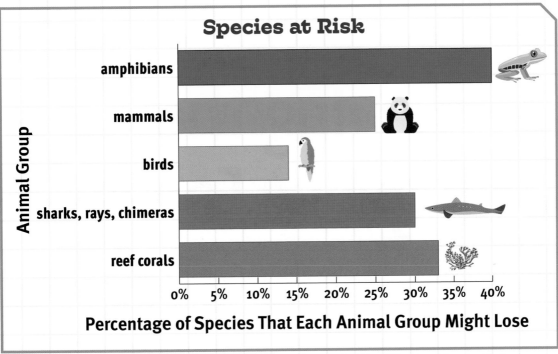

Species at Risk

Animal Group (vertical axis)

- amphibians
- mammals
- birds
- sharks, rays, chimeras
- reef corals

0% 5% 10% 15% 20% 25% 30% 35% 40%

Percentage of Species That Each Animal Group Might Lose

Source: The proportion of extant (i.e., excluding extinct) species in the IUCN Red List of Threatened Species. Version 2019-20

Analyze It!

1. Which group of animals shown is set to lose the highest proportion of its species?

2. Which group is at risk of losing exactly a quarter of its species?

3. Researchers estimate that there are about 18,000 bird species in the world. Estimate how many bird species are at risk of going extinct.

More than 25,000 animals live in Ngorongoro.

Saving the Mangrove Forest

The Ma Kôté mangrove forest is an important ecosystem on the island of Saint Lucia, in the Caribbean. Mangroves grow in water, and many young fish live among the tree roots. The forest also protects

MAKING CHANGE
Kids worked together to clean up trash in the mangrove forest.

the coast from flooding, and the trees clean carbon dioxide from the air. But the forest was in trouble not long ago. About 10 percent of the mangroves had died! To restore the forest and help the fish, over 400 students planted more than 4,000 trees.

Kids take part in a bird-watching project.

You can help animals, too! Call a local organization and find out what they are working on, and how can you help. You can also enlist your class. For example, you might participate in a project helping scientists track biodiversity by bird-watching. Your school could grow a pollinator garden. Finally, you could write to local and national politicians to tell them why you think it is important to protect biodiversity.

True Statistics

Percentage of marine and land species that went extinct 200 million years ago: 75

Number of species with backbones that have gone extinct since the early 16th century because of human activity: 680

Amount of warming that will put 16 percent of animal and plant species at risk of extinction: 7.7°F (4.3°C)

Amount of average sea level rise since the early 20th century: 8 inches (20 cm)

Speed at which the Arctic is warming compared to the rest of Earth: 2x

Percentage by which the ocean's acidity has increased over the past 200 years: 30

Percentage of marine animal species that depend on coral reefs: 25

Number of eggs a flamingo lays in its nest: 1

Did you find the truth?

(F) Climate change will signficantly decrease mosquito populations.

(T) Climate change is making it hard for ocean life to grow shells.

Resources

Other books in this series:

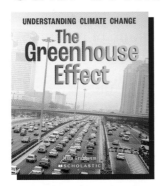  

You can also look at:

Arnold, Caroline. *A Warmer World: From Polar Bears to Butterflies, How Climate Change Affects Wildlife*. Watertown, MA: Charlesbridge, 2012.

Collard, Sneed. *Hopping Ahead of Climate Change: Snowshoe Hares, Science, and Survival*. Missoula, MT: Bucking Horse Books, 2016.

Gray, Susan H. *Ecology: The Study of Ecosystems* (A True Book). New York: Children's Press, 2012.

Green, Jen. *50 Things You Should Know About the Environment*. Lake Forest, CA: QEB Publishing, 2016.

Shea, Nicole. *Animals and Climate Change*. New York: Gareth Stevens Publishing, 2014.

Glossary

acidic (ah-SID-ick) reacting with a base to form a salt

atmosphere (AT-muhs-feer) the mixture of gases that surrounds a planet

biodiversity (BYE-oh-duh-VUR-si-tee) the condition of nature in which a wide variety of species live in a single area

climate change (KLYE-mit chaynj) global warming and other changes in the weather and weather patterns that are happening because of human activity

drought (drout) a long period of time without rain

ecosystem (EE-koh-sis-tuhm) all the living things in a place and their relationship to their environment

emissions (i-MISH-uhnz) substances released into the atmosphere

fossil fuels (FAH-suhl FYOO-uhlz) coal, oil, and natural gas, formed from the remains of prehistoric plants and animals

global warming (GLOW-buhl WAR-ming) rise in temperature around Earth due to heat from the sun trapped by greenhouse gases in the atmosphere

greenhouse gases (GREEN-hous GAS-ez) gases such as carbon dioxide and methane that contribute to the greenhouse effect

pollinators (PAWL-ih-nay-terz) animals that move pollen from one plant to another, causing plants to make seeds or fruit

Index

Page numbers in **bold** indicate illustrations.

About the Author

Katie Free is a science writer based in Austin, Texas. She discovered her love of science and nature after writing a second-grade book report about Jane Goodall. Katie has worked for *Scientific American*, *National Geographic*, *Science Friday*, *Popular Mechanics*, and Scholastic. Now she writes about science and math for kids, covering subjects from waterslides and robots to fossilized poop and giant snails.